HANDS-ON

Science

Matter and Materials

Peter Mellett

Illustrations by David Le Jars

KINGFISHER

KINGFISHER
Kingfisher Publications Plc
New Penderel House,
283–288 High Holborn,
London WC1V 7HZ
www.kingfisherpub.com

Produced for Kingfisher by PAGE*One*
Cairn House, Elgiva Lane, Chesham
Buckinghamshire HP5 2JD

For PAGE*One*
Creative Director Bob Gordon
Project Editor Miriam Richardson
Designers Monica Bratt, Tim Stansfield

Illustrator David Le Jars

For Kingfisher
Managing Editor Clive Wilson
Production Manager Oonagh Phelan
DTP Co-ordinator Nicky Studdart

First published by Kingfisher Publications Plc 2001
10 9 8 7 6 5 4 3 2 1
TS/0504/TIMS/GRS/150 MA/F

A CIP catalogue record for this book is available from the British Library

ISBN 0 7534 0273 4

Printed in China

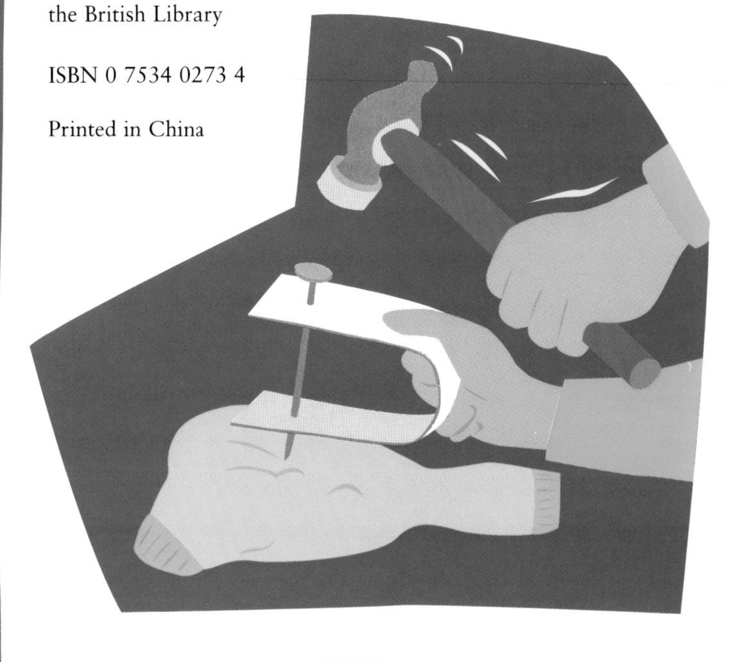

CONTENTS

Getting started

The world we live in is made from matter. What exactly is this? Matter includes anything that has mass and takes up space. Our bodies, the air we breathe and the water we drink are all examples of matter. The different types of matter that we use to make things from are called materials.

Some materials, like rocks, soil, air, water and wood, are natural.

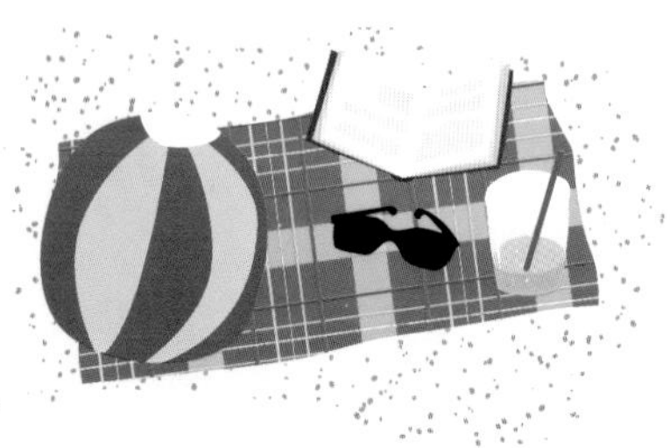

Other materials, like metals, glass, plastics and paper, are manufactured, or made by people.

This book shows you how different sorts of matter and materials behave. It will help you understand how different materials are tested and chosen before they are used in manufacturing or building.

The right stuff

You'll need a few everyday things like string, rubber bands, a plastic bottle, and some other items you can find in the kitchen.

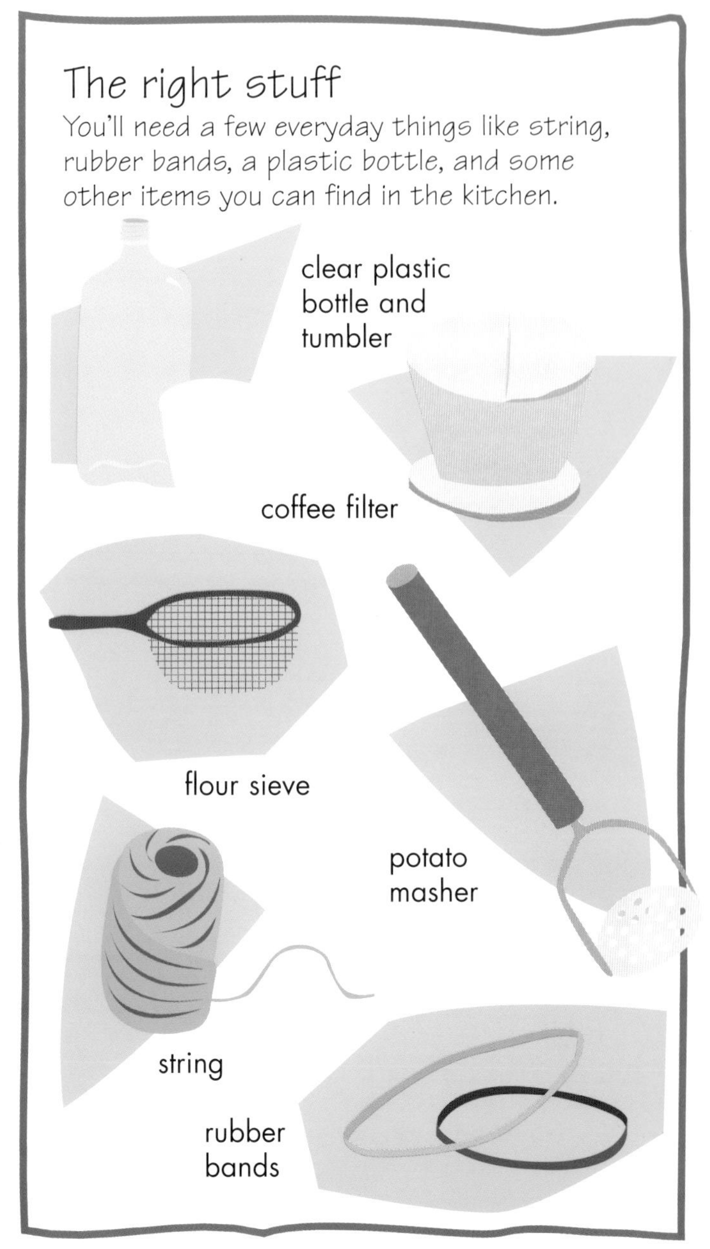

Getting organised

Carry out your experiments on a firm table. But don't forget to cover it first with newspaper to protect its surface.

If you need to pour water, put a shallow tray underneath to catch any spills.

When using a hammer, first put down an old chopping board on a firm surface like a table or the floor.

Clock symbol

The clock symbol at the start of each experiment shows you approximately how many minutes the activity should take. All the experiments take between 5 and 40 minutes. If you are using glue, allow extra time for drying.

Warning

Some activities involve heat or flames, or the use of a hammer. Ask an adult for help with these, and with any other activities where you see this warning symbol.

Don't touch your face or rub your eyes, especially if you are using materials such as salt, washing soda or soil.

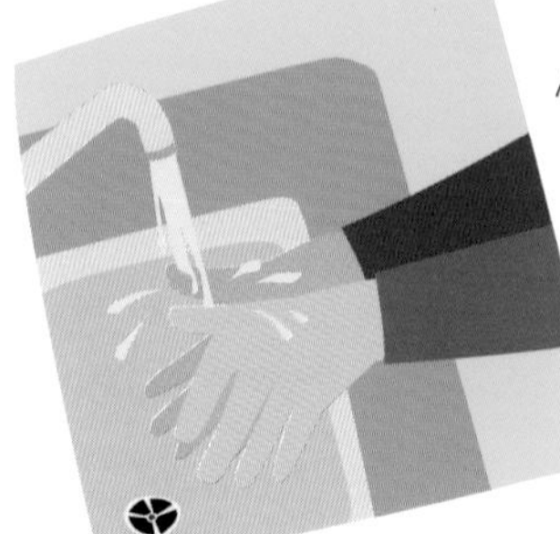

Always wash your hands and scrub your nails thoroughly after you have finished working.

Having problems?

Don't give up if at first you have problems with some of the activities. Even Einstein had his bad days!

If things don't seem to be working, read through each step of the activity again and then have another go.

If you get really stuck, remember that adults at home can help you with explanations. School teachers can also help if you show this book to them.

Stuck for words?

If you come across a word you don't understand, or you just want to find out a bit more, have a look in the Glossary on pages 38 and 39.

Denting and squeezing

Different materials have different properties – for example, the materials your clothes are made of are soft and stretchy, but materials used for building, such as concrete and brick, are hard and strong. Scientists test materials to measure and compare their properties. In this way, they can choose the best materials for making or manufacturing different things.

Hard or soft?

Trying to dent or scratch a material helps to show how hard it is. Ask an adult for permission to use the hammer, and to help you with this activity.

YOU WILL NEED 20

- A 10CM NAIL
- A HAMMER
- AN OLD SOCK
- CARDBOARD 5CM X 25CM
- A MAGNIFYING GLASS
- AN OLD KITCHEN BOARD
- SAMPLES OF SOLID MATERIALS SUCH AS WOOD, A CLAY FLOWERPOT, PLASTIC AND METAL SPOONS, A PIECE OF STONE, A RUBBER

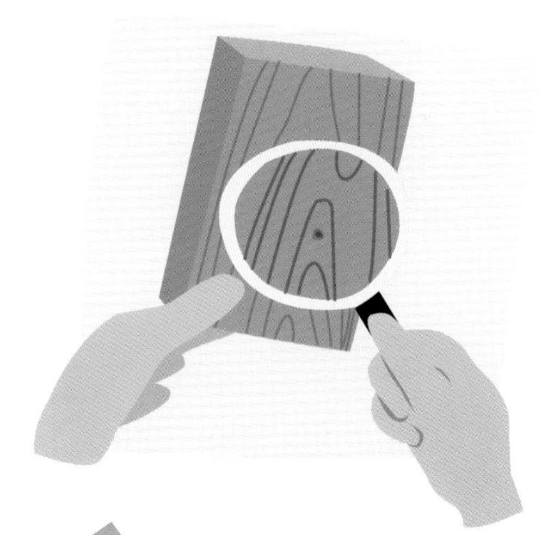

1 Make a hole 2cm from each end of the cardboard strip. Bend the strip into a 'U' shape and push the nail through the holes, making sure it is held firmly.

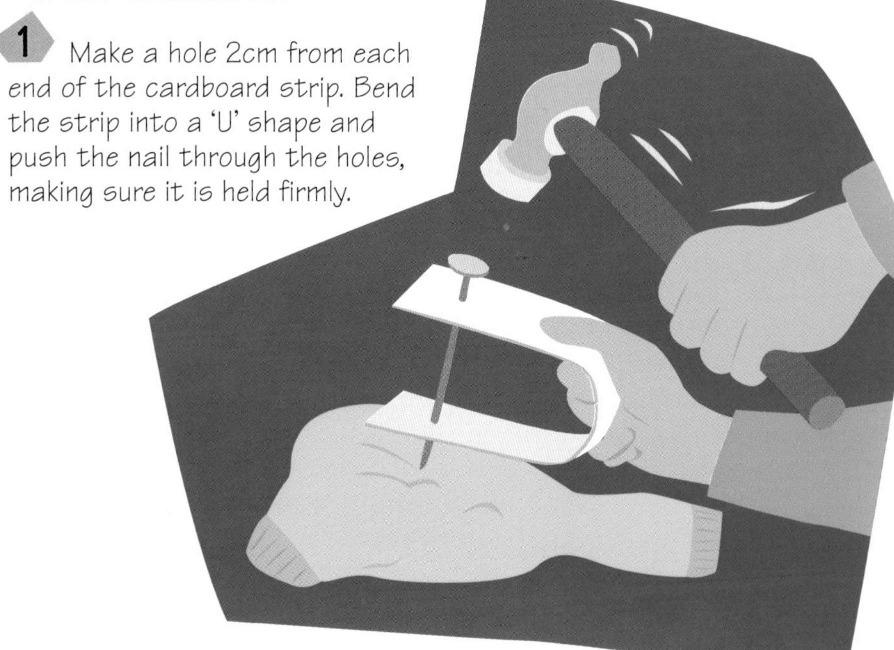

2 Put one of your samples inside the sock on top of the board. Place the point of the nail on top of the sample. Hit the head of the nail with a firm blow. The hammer should drop no more than 15cm.

3 Take the sample out of the sock. Use the magnifying glass to see whether the nail has dented or scratched the material. Now try doing steps 2 and 3 with the other sample materials and see what happens to them.

What's going on?

Different materials react in different ways to being hit hard. Some, like stone and pottery, are so hard they can't be dented at all. But they are brittle and sometimes shatter. Others, like metals, are hard but not brittle. The nail will leave a small scratch in them. Materials like wood are softer, and the nail will actually make a hole in them. Plastics can be soft, hard, tough or brittle.

Squeezing materials

Place your samples on the board and squeeze them, one at a time, under the potato masher. See what happens to each material as you steadily increase the force.

What can you feel as you squeeze each material?

YOU WILL NEED
- A POTATO MASHER
- SMALL PIECES OF PENCIL ERASER, DRIED PASTA, MODELLING CLAY

5

What's going on?

Squeezing a material tests how well it stands up to a force called compression. Elastic materials like the pieces of pencil eraser spring back when the force stops. Brittle substances like dried pasta shatter instead. Modelling clay isn't elastic like the pencil eraser. The force breaks it up so that it is squeezed through the holes of the potato masher.

FLASHBACK

The first metal

Six and a half thousand years ago, people in Egypt were the first to discover metal – in the form of copper. Like most metals, copper is found in stony materials called ores. The Egyptians blasted air into a furnace to make charcoal burn white-hot. This freed the copper so they could collect it and use it to make things with.

MATERIALS IN YOUR LIFE
How much of our world today is natural? The natural materials you are most likely to see are wood, stone, cotton and wool. Most materials are manufactured. Plastics are made from crude oil. Metals like iron, steel, copper and aluminium come from rocky ores.

Stretching and snapping

The extent to which a material can be pulled and stretched is called its tensile strength. Materials with a great tensile strength are chosen by engineers to do certain jobs. The steel cable of a crane, for example, has a high tensile strength and is able to support a very heavy load.

Threads and wires

Compare the tensile strength of three different materials. Remember to repeat the activity using a different thread each time!

YOU WILL NEED 20

- A 2-LITRE PLASTIC BOTTLE
- A BROOM HANDLE
- TWO KITCHEN STOOLS OR CHAIRS
- A MEASURING JUG CONTAINING A LITRE OF WATER
- A MARKER PEN
- THREE THREADS OF THE SAME THICKNESS, EG REAL WOOL, DENTAL FLOSS (NYLON), FUSE WIRE (COPPER)

1 Lie the broom flat on top of the two stools, as shown in the picture.

2 Pour a litre of water into the bottle. Do this slowly so you can mark the water level on the bottle for every 100ml you pour. 100ml of water has a mass of 100g, so label the marks 100g, 200g, 300g, and so on.

3 Tip the water out. Tie the end of one thread around the neck of the bottle and the other end around the broom handle. The bottle should hang a little above the floor.

4 Support the bottle with one hand and slowly pour water in. After every 100ml, replace the cap, then let your hand go. When the thread breaks, note the water level in the bottle. Repeat steps 3 and 4 with the other threads. Which one lasts the longest without breaking?

What's going on?

Gravity pulls downwards on the water in the bottle. This creates a pulling force called tension in the thread,which causes it to stretch and then snap. How quickly this happens depends on how thick the thread is and what it is made from. Wool, a natural fibre, isn't very strong. Dental floss, made from a type of plastic called nylon, and copper fuse wire both have a much greater tensile strength than wool.

Testing thin sheets

Cut out strips of each material 1cm wide by 15cm long. Wrap a strip tightly around the clothes peg and hold it firmly. Squeeze the peg harder and harder until the material breaks. Do this with each strip.

Which materials snap most easily?

YOU WILL NEED

15

- THIN SHEETS OF MATERIAL, EG A CRISP PACKET, CLINGFILM, NEWSPAPER AND A PAPER TOWEL
- A CLOTHES PEG
- A PAIR OF SCISSORS

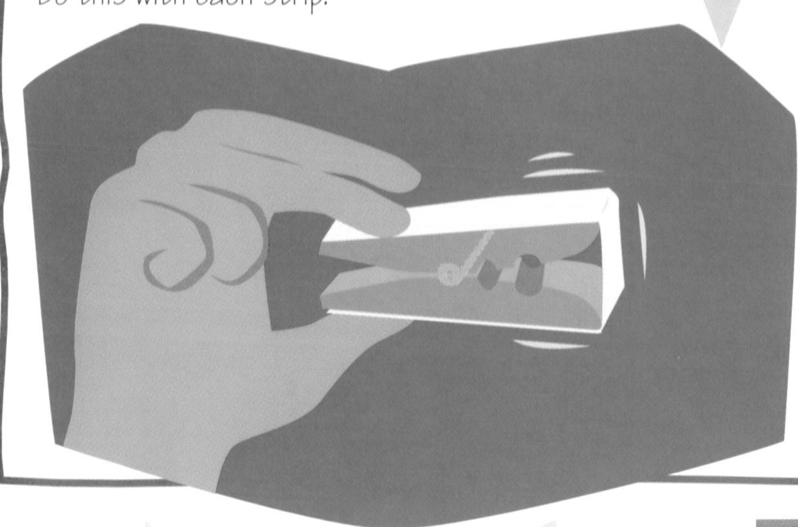

What's going on?

Some materials are more elastic than others, which means they stretch further before they break. Paper materials are made from particles called fibres, which break apart quite easily. Plastic materials like clingfilm are made from particles called molecules. These hold strongly together, stretching before they part.

FLASHBACK

Poly bags

Polythene is made from a gas called ethene. It was invented in the UK in 1933 and was used to insulate cables in aircraft radar sets. During the 1950s, polythene replaced paper for wrapping food and for making carrier bags. Nowadays, we use poly bags like this for carrying shopping, but if you overload them, they'll eventually break!

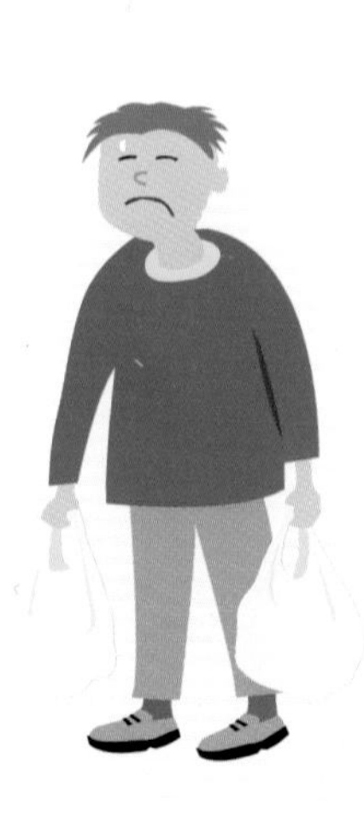

CRANES AND CABLES
The cables of a crane are made from steel. They can support huge loads without breaking. The tensile strength of steel is four times greater than copper and ten times greater than nylon.

Soil

Soil is one of the most important materials in the world. Nearly all plants need soil to grow and most animals depend on plants as the source of their food. If there was no soil, there would be almost no life on the land. There are many different sorts of soil, but all of them are a mixture of sand, clay and the rotted remains of dead plants called humus.

Testing soil

Find out what your local soil is like – how much water does it absorb and how well does water drain through it?

YOU WILL NEED 20

- DRY SOIL
- A 500ML PLASTIC DRINKS BOTTLE
- A SPOON
- SCISSORS
- COTTON WOOL
- A TABLESPOON
- A MEASURING JUG OF WATER

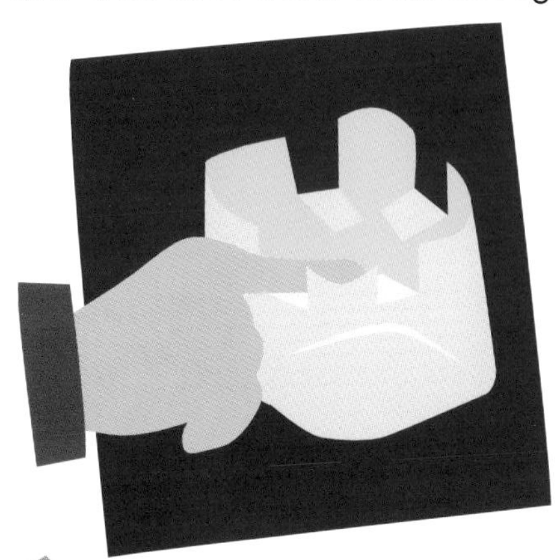

1 Cut the bottle in half. Make two cuts down each side of the lower part of the bottle. Fold the cut parts inwards to make four tabs.

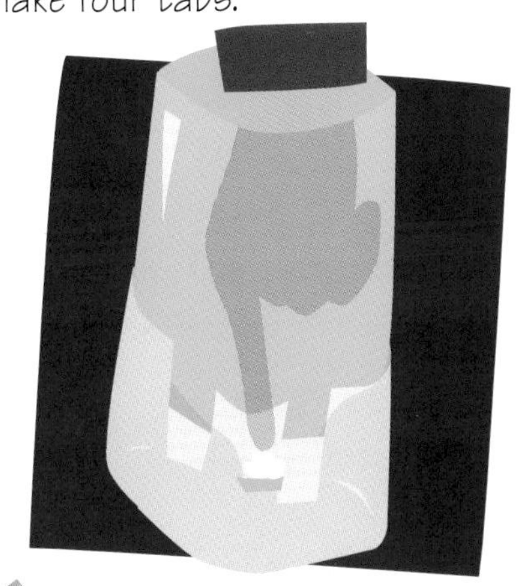

2 Turn the top half of the bottle upside down to make a funnel. Now push it down into the bottom half of the bottle so that the tabs grip the bottle neck. Push a ball of cotton wool into the bottle neck.

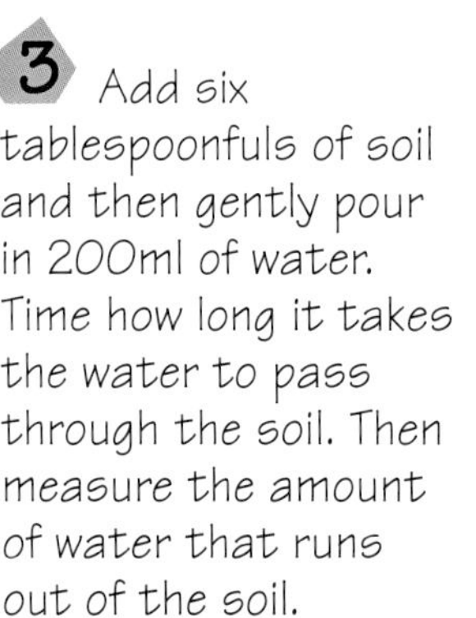

3 Add six tablespoonfuls of soil and then gently pour in 200ml of water. Time how long it takes the water to pass through the soil. Then measure the amount of water that runs out of the soil.

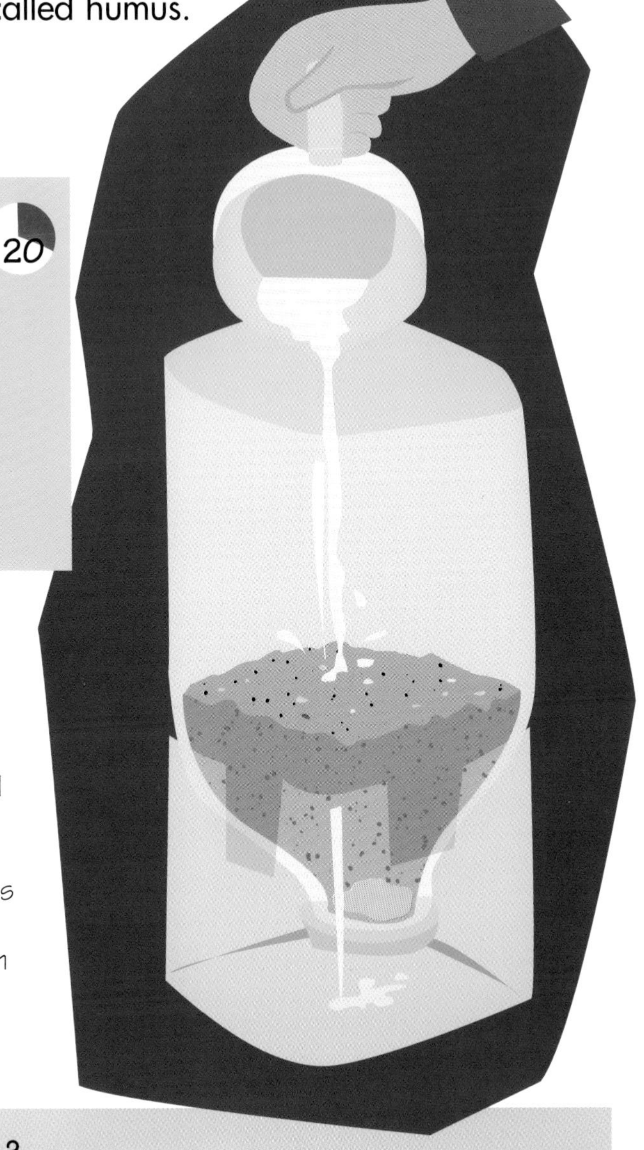

What's going on?

Water drains through soil by trickling through the spaces between the soil particles. Not all of the added water drains out because some is absorbed by the clay and humus in the soil. The sandier the soil, the more water will drain out. This is because sand does not absorb water. Clay particles are hundreds of times smaller than sand grains. They block the spaces between sand and humus and slow the downward movement of water. Water cannot pass at all through some clay soils. You can tell from the amount of water that drains through whether your soil is full of sand or clay.

What's in your soil?

Quarter fill the bottle with soil and then fill two thirds with water. Screw on the cap and shake hard. Let the bottle stand and watch the different layers form as the soil settles.

YOU WILL NEED 10
- A PLASTIC BOTTLE WITH CAP
- DRY SOIL
- WATER

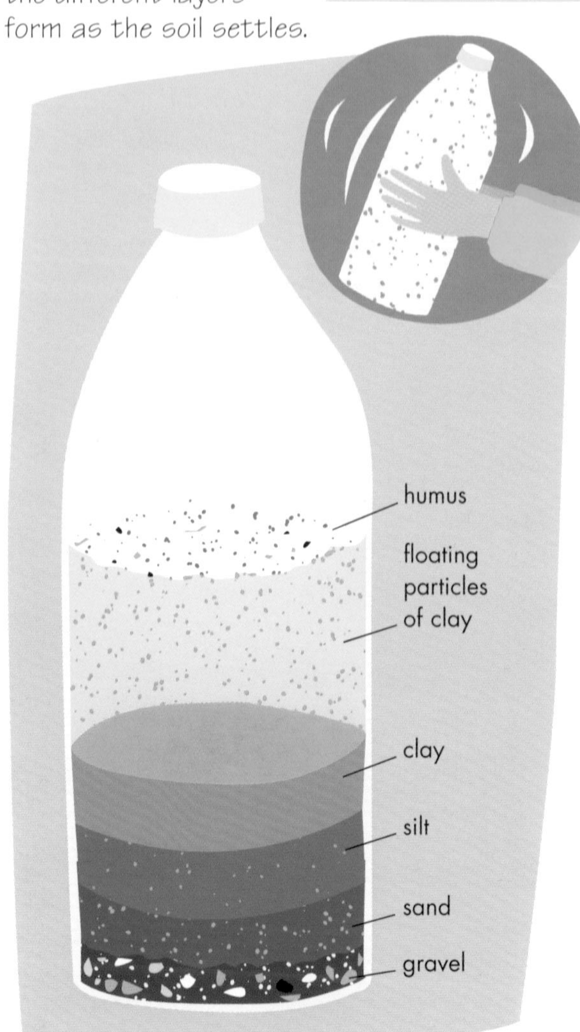

What's going on?

Large grains of sand and gravel are the first to settle to the bottom. The next layer up is fine silty sand, followed by clay particles. Floating above the clay layer are tiny clay particles too small to settle to the bottom. You may also see humus floating on the surface. By testing different soil samples, you can see how soil varies from place to place.

Hardness of rocks

Rub each rock sample on the emery paper to see how easily it crumbles into powder. After rubbing, look at each surface to see how smooth or rough it is.

YOU WILL NEED 10
- SAMPLES OF DIFFERENT ROCKS, EG CHALK, SANDSTONE, GRANITE
- COARSE EMERY PAPER

How do you think sand is made in nature?

What's going on?

Emery paper is coated with very hard and abrasive (rough) particles. They cut easily into soft rock and break it down into a sandy powder. The weather has the same effect on rocks, but it takes millions of years to change large boulders into grains of sand.

SOIL POCKETS
All over the world, mountain plants grow in tiny pockets of soil hidden in cracks between the rocks. Soil contains chemicals that all plants need to grow.

Moving heat

Heat moves through solids by a process called conduction. Some materials, like metals, allow heat to pass easily. They are good conductors of heat. Other materials, like paper and plastics, do not allow heat to pass easily. They are called insulators and are poor conductors of heat. We use insulating materials to keep things warm.

Heat loss

Hot drinks cool down because heat moves from the hot liquid to the cooler air outside.

Keeping warm

Find out which insulating material keeps a hot drink hottest for the longest.

YOU WILL NEED
- FOUR CHINA MUGS
- A POLYTHENE BAG WITH TIE HANDLES
- FOUR THIN RUBBER BANDS
- NEWSPAPER
- COTTON WOOL
- HAND-HOT WATER
- A CLOCK

20

1 Wrap layers of newspaper around a mug and hold in place with a rubber band. Cover another mug with cotton wool. Place the third mug upright in an open polythene bag and leave the fourth uncovered.

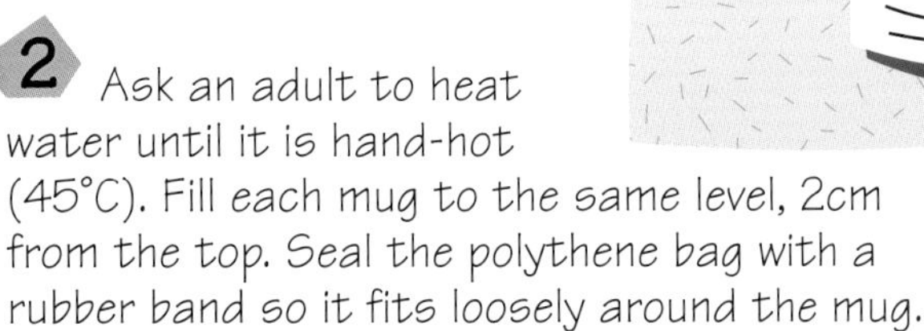

2 Ask an adult to heat water until it is hand-hot (45°C). Fill each mug to the same level, 2cm from the top. Seal the polythene bag with a rubber band so it fits loosely around the mug.

What's going on?

You should find that the water in the mug sealed inside the polythene bag has stayed the hottest, while the water in the uncovered mug is the coolest. Air is a good insulator, so long as it doesn't move around too much. The polythene bag holds a layer of air around the mug that stops heat escaping. Cotton wool contains air trapped in its fibres. Newspaper also contains air, although less than cotton wool. Most insulating materials rely on trapped air to stop heat flowing away.

3 After 15 minutes, use a finger to test the water in each mug. Arrange the four mugs in order, from the hottest down to the coolest.

Testing heat conduction

Stick a bead to the handle end of each spoon with a blob of butter or margarine. Stand the bowl on the newspaper and arrange the spoons so their handles stick out around the rim. Ask an adult to pour freshly boiled water into the bowl. Time how long it takes for the bead to drop from each spoon.

YOU WILL NEED

20

- BUTTER OR MARGARINE
- A METAL, PLASTIC AND WOODEN SPOON
- A HEAT-PROOF GLASS BOWL
- THREE SMALL PLASTIC BEADS
- BOILED WATER (ASK AN ADULT)
- NEWSPAPER

What makes the beads drop off?

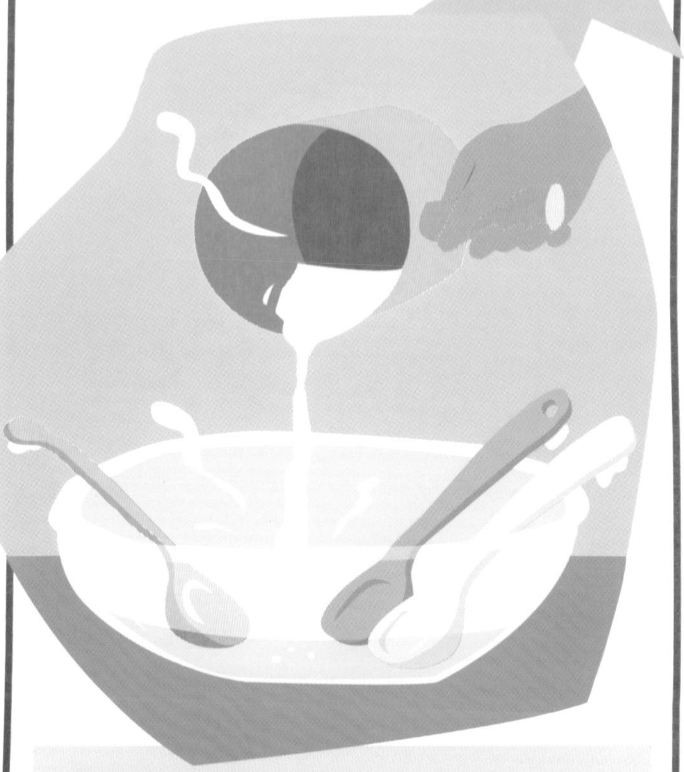

What's going on?

Conduction carries heat up the handle of each spoon, which causes the butter to melt and the bead to fall off. Metal is a better conductor than wood or plastic. The end of the metal spoon becomes hot quickest and the bead drops off this spoon first. It takes longest for the bead to fall from the wooden spoon because wood contains air and is a poor conductor of heat.

FLASHBACK

Heat is a form of energy

Two hundred years ago, scientists thought that heat was an invisible fluid. But in 1851, William Thomson introduced the modern idea that heating something increases the energy of its particles and makes them move about more rapidly.

WINTER WARMTH

Have you ever wondered why birds puff themselves up in cold weather? Under their feathers are fluffy fibres that trap layers of air. These insulating layers cut down heat loss and keep the birds warm.

Solids, liquids and gases

Our world is made from millions of different materials, but all this matter exists in just three main forms – as solids, as liquids or as gases. Solids such as bricks and ice cubes are hard and have a fixed shape. Liquids such as water are runny and do not have a fixed shape. They have a flat surface and fill the bottom of a container. Gases spread out in all directions, so they are often kept in a closed container.

Feel the difference

See what happens when you try to squeeze a gas (air), a liquid (water) and a solid (ice). You'll need extra time to make the ice in step 3!

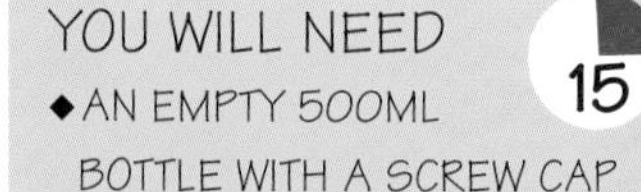

YOU WILL NEED

15

- AN EMPTY 500ML BOTTLE WITH A SCREW CAP
- WATER
- A LONG BALLOON
- A FREEZER

1 Take the empty bottle, with its cap screwed on tightly, and squeeze it in your hand. What happens to the bottle?

2 Now unscrew the cap and, over a sink, fill the bottle with water until it overflows. Screw the cap on tightly and try squeezing the bottle again. Can you still squash it easily?

3 Fill a balloon with water (over a sink!) and then tie it up. Squeeze the balloon and feel how the water moves around inside. Place the balloon in a freezer for one hour. Now see if you can move the water around inside the balloon.

What's going on?

Air is a gas and is compressible, which means it can be squeezed into a smaller space. Water is a liquid and is not compressible, which is why you can't squeeze the bottle filled with water. Liquids and gases are called fluids because they can flow from one place to another. When the temperature falls below 0°C, water freezes to make solid ice. Solids cannot flow and they are not compressible.

Gases have mass

Tie a piece of string to each end of the wooden rod. Tie the other end of each piece of string to the ring-pull on the cans. Hang the rod by its centre from underneath the stool so that the cans are evenly balanced. Ask an adult to gently pull the ring on one can to open it slightly, then let the cans hang in balance.

How does the balance change over the next half hour?

YOU WILL NEED 15

- TWO CANS OF FIZZY DRINK WITH RING-PULLS
- A THIN WOODEN ROD 30CM LONG
- A KITCHEN STOOL OR CHAIR
- STRING

What's going on?

You'll see that the balance of the cans is disturbed and that the open can rises slightly. This is because fizzy drinks contain a gas called carbon dioxide dissolved in flavoured water. Once the can has been opened, the carbon dioxide escapes slowly from the liquid, causing the mass of the liquid to decrease. This means the contents of the open can weigh less than they did when it was closed.

FLASHBACK

John Dalton was an English chemist who lived from 1766 to 1844. He said that matter is made up from invisible particles. In solid materials, the particles are fixed together, but in liquids they can slide around past each other. The particles in gases are far apart and travel at great speed.

INFLATABLE DINGHY
Gases like air are made up from particles that are far apart. But to inflate this rescue dinghy, air has been pumped in under pressure, which squeezes the air particles closely together. This makes the dinghy firm and buoyant in the water.

Mixing materials

Most materials are not one single pure substance. They are usually made up of different substances mixed together in different ways. For example, pastry dough is a mixture of flour, fat and water, while fizzy drinks consist of water, sugar, flavourings and carbon dioxide gas. The right ingredients must be chosen to make up each different mixture.

Which of the grains disappear in the water?

Solutions

We can mix water with sugar or salt to make something called a solution. The solution behaves differently to ordinary water.

YOU WILL NEED 20

- WARM WATER
- SOLID MATERIALS SUCH AS SUGAR, SALT AND SAND
- FOUR CLEAR PLASTIC CUPS
- A TEASPOON
- A MAGNIFYING GLASS
- A FREEZER

1 Place a few grains of each solid on the table. Look at them through the magnifying glass. Can you see a difference in their shape and size? The grains of salt and sugar have straight sides – they are called crystals.

2 Half fill one of the cups with warm water. Add a pinch of sugar and watch what happens to each grain. Then add a heaped spoonful of sugar and stir the mixture. Notice how the grains dissolve and disappear.

What's going on?

Sugar and salt crystals dissolve (break down) when they are mixed with water. We call the result a sugar solution (or salt solution). When the crystals dissolve, they break into particles that are too small to see. These particles spread evenly through the water. Dissolved substances make freezing happen at a lower temperature than with pure liquids, so the sugar and salt solutions will take longer to freeze than the pure water. Unlike sugar and salt, sand is insoluble – it doesn't dissolve.

3 Half fill another cup with water. Place this and the cup of sugary water in the freezer for two or three hours. Try to look at them every 15 minutes to see what's happening. Now repeat steps 2 and 3 with the salt, and then the sand.

Investigating cake mixture

Ask an adult to help you collect all the ingredients and equipment needed to bake some small cakes. Watch how the ingredients change as you mix them together. Then see the mixture change again once it has been baked in the oven.

YOU WILL NEED
- INGREDIENTS AND KITCHEN EQUIPMENT FOR BAKING CAKES
- AN OVEN

40

How do you change soggy cake mixture into cakes?

What's going on?

It's amazing how different a baked cake looks and tastes from the original raw ingredients. Cake mixture usually contains flour, eggs, sugar and fat. While it cooks in the oven, heat makes the mixture expand and changes its colour, texture and taste.

FLASHBACK

16th-century alchemist in his laboratory

Alchemists were early chemists who worked more than 400 years ago. They boiled, melted and dissolved things a bit like modern chemists do today. They believed that if they mixed the right ingredients together, they could change cheap metals into gold. They didn't understand that mixtures contain different substances arranged in a certain way.

SAWING UP CHIPBOARD
Wood is a great material for making things. But it has a disadvantage – a plank of wood can't be any wider than the tree trunk from which it is cut. So manufacturers produce chipboard – wood chippings and sawdust bonded together by glue. It comes in big sheets and is easy to cut into shape.

Expansion and contraction

When solids, liquids and gases are heated, they take in energy and their temperature gets higher. As this happens, the substance expands. It takes up more space and we say that its volume increases. As the substance gets cooler, it loses energy and its temperature decreases. This time the volume decreases and the substance contracts, or gets smaller.

Hot and cold air

This activity helps you to see how air – an invisible gas – expands and contracts when it is heated and cooled. Ask an adult to help you with the glass and the hot water.

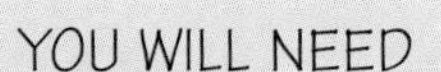

YOU WILL NEED

15

- A SMALL, STRONG GLASS BOTTLE, EG EMPTY KETCHUP BOTTLE
- A DRINKING STRAW
- MODELLING CLAY
- A TEA TOWEL
- HOT WATER (ASK AN ADULT)
- A COLD, WET CLOTH
- A BOWL OF WATER

1 Gently wrap a ball of modelling clay around the straw near to one end. Push the clay firmly into the neck of the bottle to make an airtight seal.

2 Ask an adult to soak the tea towel in hot water and wrap it all around the bottle.

3 Turn the wrapped bottle upside down and dip the end of the straw under the water in the bowl. What do you notice?

4 Keep the end of the straw under the surface of the water. Unwrap the hot cloth, then wrap the cold cloth around the bottle. Now watch what happens to the water!

What's going on?

By heating the bottle with the warm cloth, you are heating the air inside it. Heat energy makes the tiny particles of air move around faster and they take up more space. As a result, the air expands and bubbles out of the straw. Cooling the bottle has the opposite effect. The particles slow down and take up less space. The air contracts and water enters the bottle.

Heating water

Use the same equipment as before. This time, fill the bottle to the top with cold water before you fit the straw. Make sure some water rises about halfway up the straw, and mark the position with a pencil. Now stand the bottle in a bowl of hot water and watch the water level in the straw.

YOU WILL NEED 10

- BOTTLE, STRAW AND MODELLING CLAY AS BEFORE
- A BOWL OF HOT WATER (ASK AN ADULT)
- A PENCIL

Do liquids expand as much as gases?

What's going on?

The particles in water move more slowly than the particles in air. Liquid particles are closer together than gas particles and must slide past each other as they move. Heating water makes the particles move faster which makes the liquid expand and rise up the straw. But the effect isn't so great as the expansion of gases.

FLASHBACK

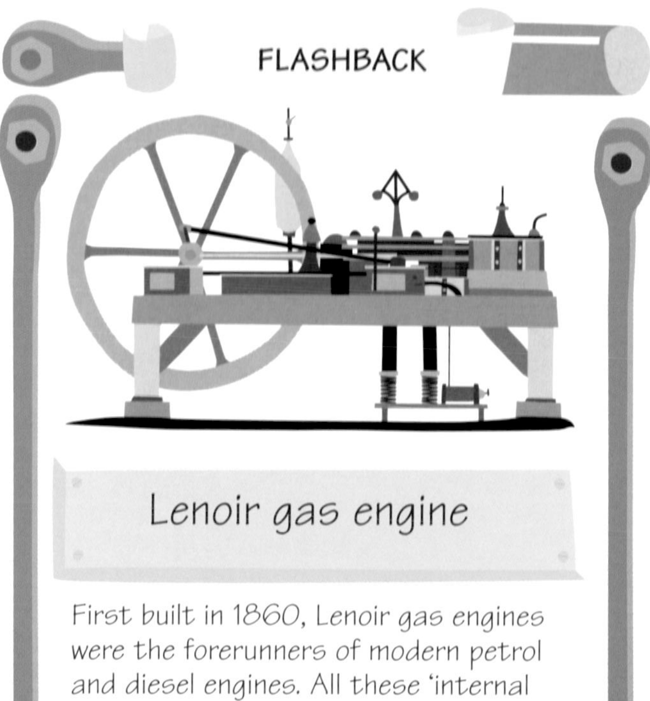

Lenoir gas engine

First built in 1860, Lenoir gas engines were the forerunners of modern petrol and diesel engines. All these 'internal combustion' engines burn a mixture of fuel and air inside a cylinder. The heat makes the gases expand, which forces a piston to move inside the cylinder. The piston is attached to a crank that works like the pedals of a bicycle and spins a wheel.

HOW THERMOMETERS WORK
When someone takes your temperature, a liquid metal (mercury) inside the thermometer responds to the heat in your mouth by expanding. As it does this, the mercury moves along a very thin tube that is marked by a temperature scale.

Heating substances

When you heat a substance, its temperature increases. This rise in temperature causes many substances to change their appearance. For example, water bubbles when it boils and bread changes into toast. When heating stops, the temperature falls again. Water stops bubbling, so we say that the change is only temporary. Toast, on the other hand, does not change back into bread when it cools. Heat here has caused a permanent change.

Gentle heating

Some substances change when the temperature rises only slightly. When you do this activity, don't touch the lamp bulb, as it will get hot.

YOU WILL NEED 20

- A BLOB OF BUTTER
- A PIECE OF CHOCOLATE
- A PIECE OF CANDLE WAX
- SUGAR
- ALUMINIUM FOIL
- SCISSORS
- AN ADJUSTABLE DESK LAMP
- A DRINKING STRAW

1 Cut out four 10cm squares of aluminium foil. Fold up the edges and pinch the corners to make four small boxes with an open top and a flat bottom.

2 Put a small amount of each substance into the aluminium boxes so that each box contains something different.

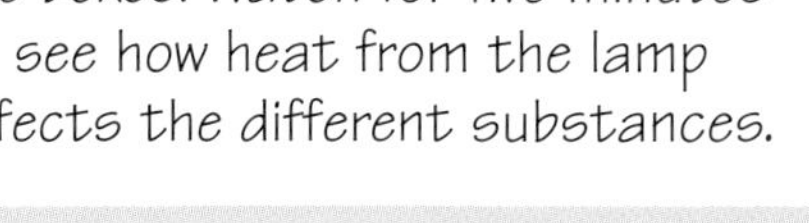

3 Ask an adult to switch on the lamp and point it straight downwards, about 5cm above the boxes. Watch for five minutes to see how heat from the lamp affects the different substances.

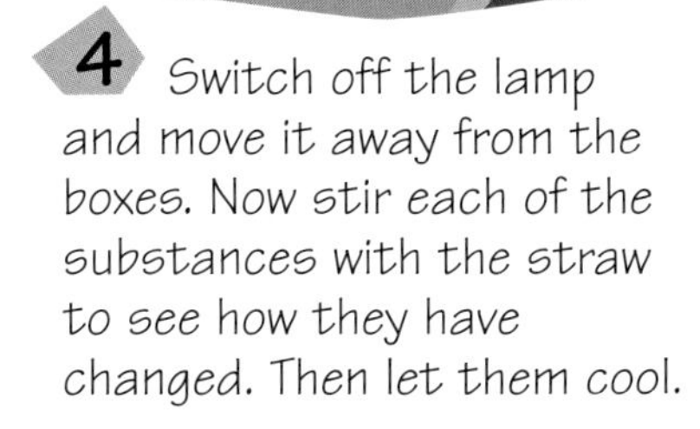

4 Switch off the lamp and move it away from the boxes. Now stir each of the substances with the straw to see how they have changed. Then let them cool.

What's going on?

The lamp raises the temperature to about 75°C and gently heats the four substances. Remember that water boils when we heat it to 100°C. The butter, chocolate and candle wax all become liquids when they are gently heated in this way. We say that they have melted. When they cool down again, they change back into solids. So melting is a temporary change. Sugar is not affected by the heat from an electric lamp and so does not change at all.

Stronger heating

Some substances need stronger heat to make them change. Set the oven to 200°C. Put a little sugar, salt and egg into each of the aluminium boxes and place them on a baking sheet. Ask an adult to put them in the oven and then get them out after 15 minutes. Which substances look different?

YOU WILL NEED 15
- ALUMINIUM BOXES (AS MADE EARLIER)
- SUGAR
- SALT
- RAW EGG
- A BAKING SHEET

What's going on?

Sugar melts at oven temperature and then starts to change into a brown, sticky substance called caramel. When caramel cools, it becomes solid. So it is a permanent change. Egg bakes inside an oven. This change is also permanent. This is because particles in the sugar and egg break apart and join up in a new way. Salt is not affected. It must be heated to over 850°C before it melts. When cool, it becomes solid salt again.

High-temperature heating

Place some sugar on an old teaspoon. Ask an adult to light the nightlight and hold the spoon over it for a while to heat it. What do you see?

YOU WILL NEED 15
- AN OLD TEASPOON
- SUGAR
- MATCHES (ASK AN ADULT)
- A NIGHTLIGHT IN A SAUCER OF WATER

What makes the sugar turn black?

What's going on?

Sugar is made up from carbon, hydrogen and oxygen. When heated to about 500°C, it breaks down into black carbon, which is what you see on the spoon, and steam, which rises. When this happens, we say the sugar decomposes. It is a permanent change.

SHAPING GLASS
Glass gradually gets softer as the temperature rises. To shape glass, workers blow down an iron pipe to make the bubbles of glass expand. They cut open the bubbles to produce beautiful jugs and ornaments.

Changing state

Matter can exist in three states – as solids, liquids or gases. When we heat a substance, it sometimes changes its state. Heat can make a solid melt to form a liquid or a liquid boil to form a gas. These changes of state are temporary because cooling reverses the changes. Gases condense into liquids and liquids freeze and become solid again.

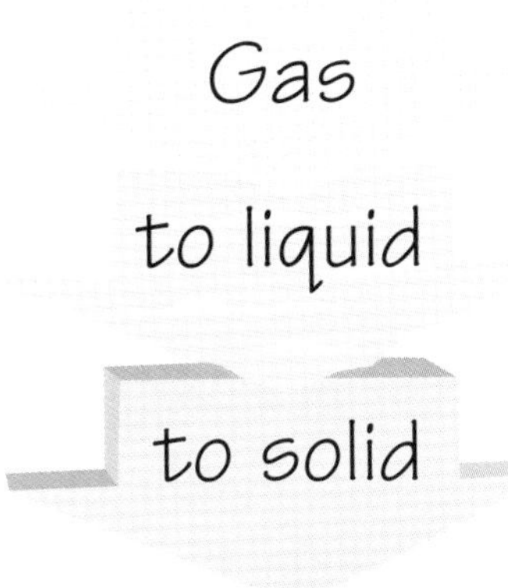

YOU WILL NEED (30)
- ICE
- A TEA TOWEL
- TWO RUBBER BANDS
- A ROLLING PIN
- SALT
- A LARGE DARK-COLOURED MUG
- A SPOON

The air is full of invisible water vapour. You can use a freezing mixture to trap this gas and turn it into ice, which you can see.

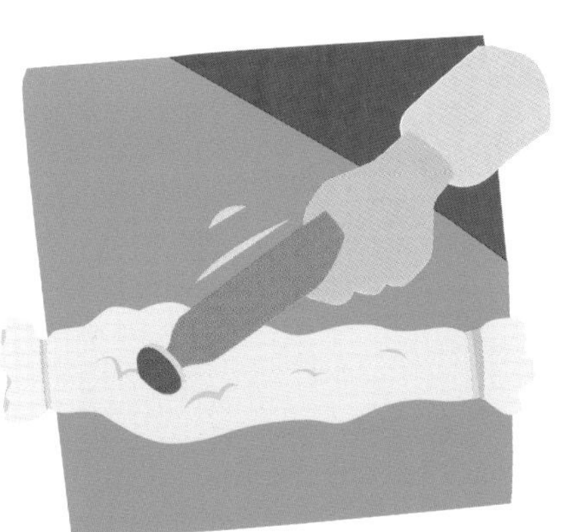

1 Place ten ice cubes along one edge of the tea towel and roll it into a sausage shape. Twist a rubber band around each end of the tea towel and place it on a firm surface. Now crush the ice with the rolling pin.

2 Half fill the mug with crushed ice. Add about a quarter of a mugful of salt. Stir the mixture and then leave the mug undisturbed for about 20 minutes.

FLASHBACK

See you later, water vapour

Over millions of years, flowing water has helped to shape the surface of the Earth. Heat from the Sun makes sea water evaporate (change into water vapour). Vapour rises into the air, cools and forms clouds of tiny water droplets. These droplets fall as rain. As rivers carry water back to the sea, they slowly carve out valleys between hills and mountains.

3 You will see that a white solid forms on the outside of the mug. It reaches up to the same level on the outside as the ice and salt inside. Scrape some of the solid into the spoon and watch it melt to form a liquid.

Boiling and evaporation

Wet two cotton handkerchiefs and wring them out. Hang one in a warm or sunny place and the other in a cool place. Check every five minutes to see how each hanky is drying.

YOU WILL NEED
- TWO COTTON HANDKERCHIEFS
- A SUNNY SPOT OR A WARM PLACE INDOORS
- A COOL PLACE
- WATER

20

Which handkerchief dries the fastest?

What's going on?

You probably won't be surprised to see that the wet handkerchief in the warm place dries faster than the one in the cool place. But why is this? As water takes in heat from the surrounding air, it changes and becomes a gas called water vapour. When this happens, we say that the water has evaporated. The higher the temperature, the quicker the rate of evaporation. So the hanky in the warm place dries faster than the one in the cool place because the water evaporates from it more quickly.

What's going on?

The temperature of ice drops even lower when salt is added. The mixture inside the mug makes the outside extremely cold. The air around us is a gas which contains water vapour dissolved in it. When this invisible vapour touches the outside of the mug, it condenses – which means it changes into liquid water. This immediately freezes into solid ice. When you scrape some of this into the spoon, it warms up and melts to form liquid water.

LIQUID STEEL
Steel melts at around 1,540°C. When this happens, it glows white hot and is nearly eight times heavier than the same amount of water. Here liquid steel is being poured into moulds to make parts for engines.

Permanent changes

As we saw on pages 20 and 21, some changes are temporary and can be easily reversed. For example, chocolate melts when it is heated but changes back to solid chocolate when cooled. Other changes are permanent and cannot be reversed. For example, boiling an egg changes it permanently. There are three main ways of making permanent changes happen – by mixing substances together, by heating them or by passing electricity through them.

Mix for a change

Mixing vinegar and bicarbonate of soda together creates carbon dioxide gas, which puts out flames. You must ask an adult to do steps 2 and 3 of this activity.

YOU WILL NEED 15

- BICARBONATE OF SODA
- A HEATPROOF PUDDING BOWL
- A SHORT CANDLE
- VINEGAR
- MODELLING CLAY
- A DESSERT SPOON

1 Use the modelling clay to fix the candle firmly in the centre of the bowl. Place five level spoonfuls of bicarbonate of soda around the candle.

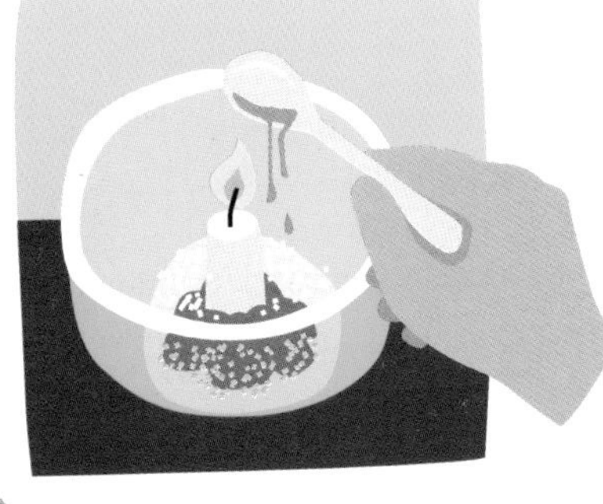

2 Ask an adult to light the candle, then to spoon vinegar down the inner side of the bowl, avoiding the flame. Watch how the liquid and powder froth as they mix.

3 Ask the adult to stop adding the vinegar when the froth is about halfway up the candle. The candle will suddenly go out. Now ask them to try to re-light it.

What's going on?

A permanent change takes place when bicarbonate of soda and vinegar are mixed together. The particles in these two substances join up in a different way to make new substances.. One of the new substances is a gas called carbon dioxide, which causes the mixture to froth. This gas is heavier than air. Although you can't see it, it fills the bowl and extinguishes, or puts out, the flame. Carbon dioxide is used in many types of fire extinguisher.

Bake a model

Mould some bakeable modelling clay into a shape to make a model. Think about how it looks and feels while you do this. Then ask an adult to follow the instructions and bake your model in the oven. Let it cool. How does it look and feel now?

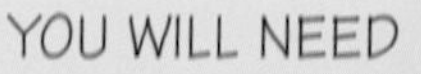

YOU WILL NEED  15

- BAKEABLE MODELLING CLAY
- AN OVEN

What changes happen when you bake the clay?

What's going on?

Modelling clay is soft and easy to squeeze. It contains long, thin particles that slide past each other when you squeeze the clay. But once the clay has been baked in the oven, permanent links form between the particles. They can no longer slide around, and this is why your baked model is now hard.

Electric effect

Get an adult to strip the ends of the wires. Ask them to connect one end of each wire to the battery and dip the other end into some salty water. Watch what happens to the wires. Do you recognise the smell?

YOU WILL NEED 20

- A 4.5- OR 6-VOLT BATTERY
- TWO INSULATED WIRES 20CM LONG
- SALT
- A CLEAR PLASTIC CUP OF WATER

What's going on?

Electricity changes part of the salt solution into a gas, which makes bubbles. This gas is called chlorine. You can smell it in swimming pools, where it is used to disinfect the water.

A CHEMICAL REACTION

These children are studying chemical reactions that make materials change permanently. The substances they start with are called reactants. After the reactants change, they make new substances called products.

Burning

The scientific name for burning is combustion. To make burning happen, you need fuel and air. Fuels can be solid, like wood and coal, or liquid, like petrol and paraffin oil. When a fuel burns, it mixes with oxygen in the air and a permanent change takes place, which gives out heat. Before they can burn quickly, solid and liquid fuels must change into gases. When they burn, many common fuels produce carbon dioxide gas and water vapour.

Air and fire

Find out how flames need air to burn, and see how they use up just a part of it. Ask an adult to do this activity for you to watch.

YOU WILL NEED 20

- A CANDLE (5CM TALL)
- A METAL ROASTING TRAY
- A TALL GLASS JAR
- WATER
- MODELLING CLAY

1 Use modelling clay to stick the candle upright in the centre of the tray. Pour water into the tray until it is about 2cm deep.

2 Ask an adult to light the candle and then to place the glass jar over it. The rim of the jar must be under the water, resting on the bottom of the tray.

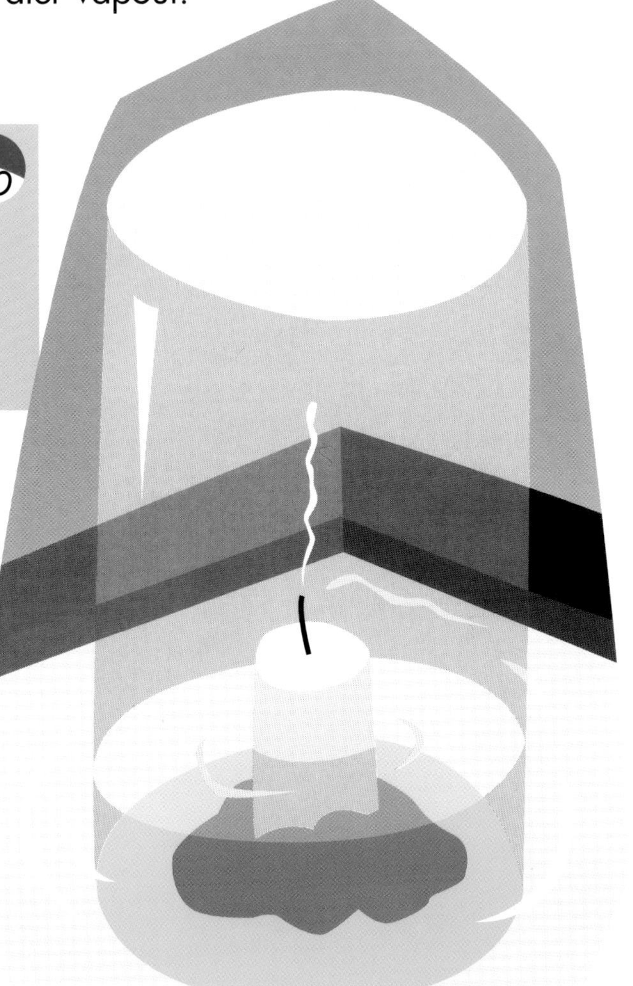

3 Watch the candle flame carefully as soon as the jar is in place. What happens to the level of the water inside the jar?

What's going on?

You'll see the water level rise inside the jar. It does this to replace the oxygen that was used up by the burning flame. Once the oxygen is used up, the flame goes out. There is still air left in the jar, but now it is mostly made up of a gas called nitrogen. Fuels cannot burn when only nitrogen is present.

FLASHBACK

Drilling for oil

Cars, lorries, ships and planes use liquid fuels made from petroleum (also called crude oil). This dark, oily liquid comes from deep under the ground. The first oil well was drilled by Edwin L. Drake in 1859 at Titusville, Pennsylvania. He struck oil at a depth of just 23 metres. Modern oil wells are up to 5,000 metres deep.

A closer look at a flame

Ask an adult to light the candle. Look carefully, but not too closely, to see the three parts of the flame.

YOU WILL NEED
- A CANDLE (5CM TALL)

5

What's going on?

The heat from the flame melts wax near the base of the wick. This melted wax soaks up the wick and into the flame. Wax on the burned part of the wick is turned by heat into a gas. This wax vapour mixes with air and burns – giving a blue part to the flame. The mixture then rises into the middle of the flame, where particles of carbon from the wax glow and give out yellow light.

FIGHTING FIRE WITH FOAM
Firefighters spray foam on to burning aircraft fuel. The foam smothers the fire and cools the burning fuel. Foam bubbles contain carbon dioxide gas and other chemicals. These prevent oxygen from helping fuel to burn.

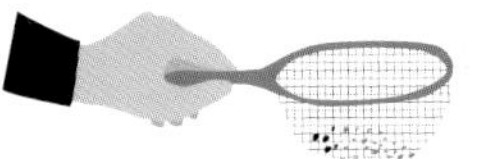

Sieving solids

We can sometimes separate mixtures by using a sieve. This only works when the mixture contains solids of different sizes. For example, we can sieve soil because it is a mixture of solids such as sand, clay and humus. The larger particles stay in the sieve while the smaller ones fall through the holes and collect underneath.

FLASHBACK

Sieving flour

Flour was once made by grinding wheat between two large revolving stones. About 120 years ago, millers in Hungary and Switzerland began to use cylinder-shaped rollers to powder the grain. Ten sieves with different size holes were stacked one above the other. They separated the milled grain into different grades of flour.

Separating soil

Use two types of sieve to sort soil into four piles of different-sized particles.

YOU WILL NEED 25

- A COLANDER
- A KITCHEN SIEVE
- FOUR LARGE PIECES OF PAPER
- STERILIZED COMPOST FROM A GARDEN CENTRE
- A MAGNIFYING GLASS

2

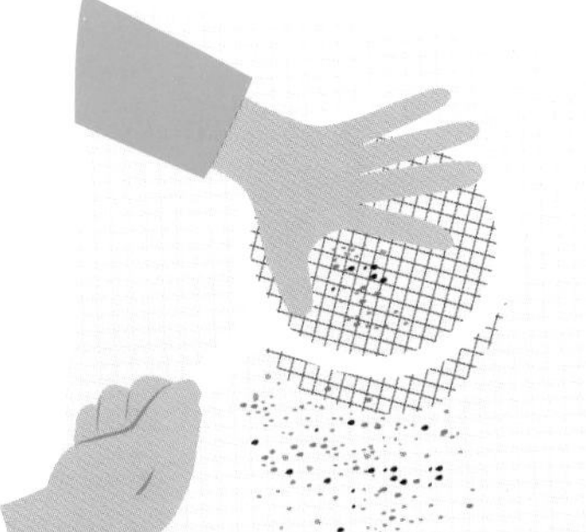

1 Number the sheets of paper 1, 2, 3 and 4. Put some compost into the colander. Hold the colander over sheet 2 and shake gently. When no more soil passes through, tip what's left in the colander on to sheet 1.

2 Take some of the particles from sheet 2 and put them into the kitchen sieve. Tap the sieve over sheet 4 until no more particles pass through. Tip what's left in the sieve on to sheet 3.

3 Gently spread out some of each pile on the sheets. Look carefully through the magnifying glass to compare the particle sizes in each of the piles.

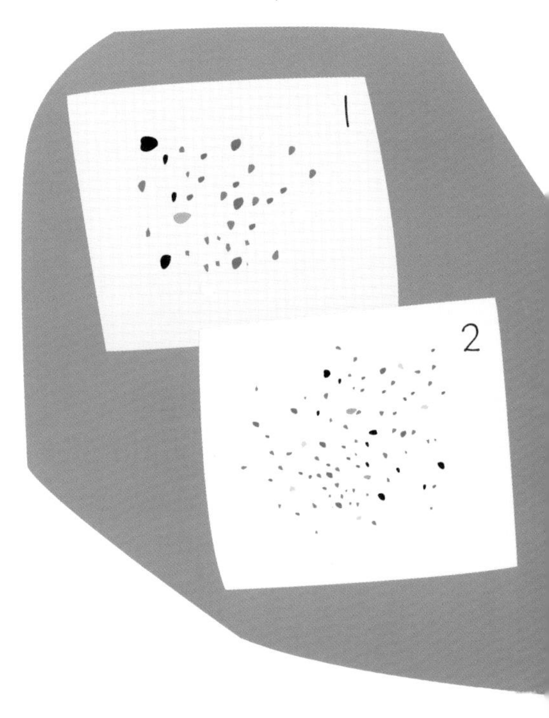

Even smaller holes

Put some soil into the tumbler. Use the rubber band to fasten the foil over the rim of the tumbler, then prick some tiny holes in the foil with the needle. Turn the tumbler upside down and then gently shake it. Look carefully at the particles that fall on to the paper.

Which soil particles pass through these tiny holes?

YOU WILL NEED 15

- WHITE PAPER
- A CLEAR PLASTIC TUMBLER
- DRY SOIL
- ALUMINIUM FOIL
- A RUBBER BAND
- A VERY FINE NEEDLE

What's going on?

The holes in the foil are much less than 1mm across. Clay particles are usually the only part of soil small enough to pass through. These particles appear as a dusty mark on the paper. You would need a powerful microscope to see a single clay particle.

What's going on?

The holes in a colander are about 4mm across. Only particles that are smaller than 4mm pass through the colander. The sieve separates the particles that are smaller than 4mm. It has holes about 1mm across. The particles that stay in the sieve are larger than 1mm but smaller than 4mm. The ones that go through are smaller than 1mm. Sheet 1 holds the largest particles and sheet 4 the smallest.

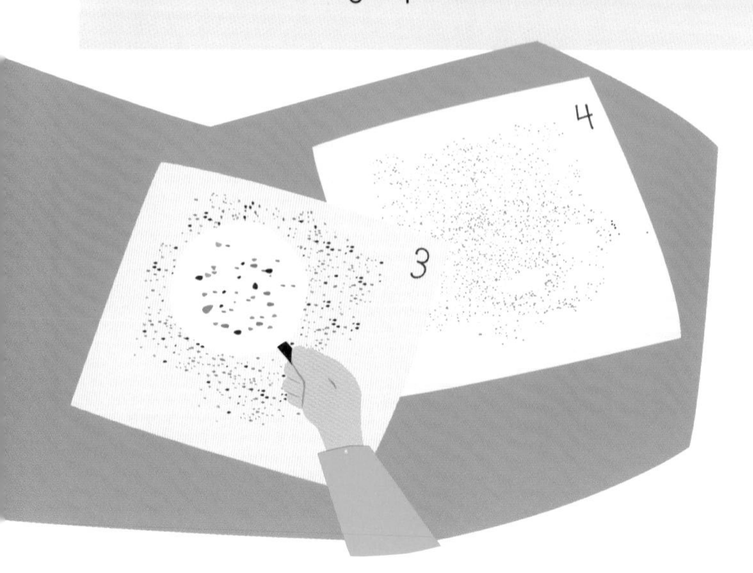

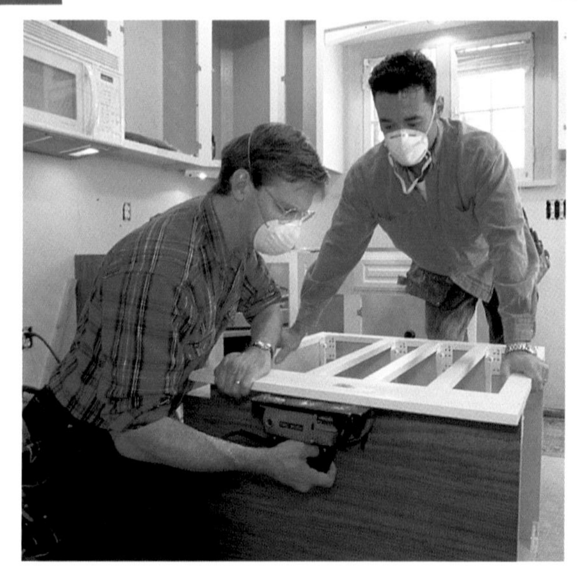

BREATHING CLEAN AIR
Wood dust from a sanding machine can harm your lungs. These men are wearing safety masks made from paper or cotton fibres. The masks work like a very fine sieve. Gaps between the fibres are large enough to allow air to pass freely, and small enough to stop wood particles from entering the mask.

Solutions and suspensions

Substances such as salt and sugar dissolve, or break down, in water. We say that they are soluble. When mixed with water, soluble substances slowly disappear as they dissolve to form a solution. Substances such as chalk and sand do not dissolve in water. We say that they are insoluble. Shaking an insoluble substance with water scatters the particles through the water and forms a mixture called a suspension.

Solution or suspension?

Add different solids to water and decide which dissolve to form a solution and which scatter to make a suspension.

YOU WILL NEED

25

- FOUR 500ML PLASTIC BOTTLES WITH CAPS
- WATER
- A TEASPOON
- A PLASTIC FUNNEL
- SUGAR, FINE SAND, INSTANT COFFEE GRANULES, PLAIN FLOUR

1 Put one spoonful of sugar into one of the bottles. Use the plastic funnel to guide the crystals into the bottle. Add the sugar slowly so that it does not block the neck of the funnel.

2 Repeat step 1, placing each of the other solids in its own bottle. Now half fill each bottle with water and screw on the cap. Shake each bottle ten times.

3 Look carefully at each bottle to see if you can still see solid particles. Decide which solids form a solution and which solids form a suspension.

What's going on?

Sugar and coffee granules dissolve in water to make a solution. All solutions are clear and you can see right through them. A sugar solution is colourless and a coffee solution is brown. Sand and flour do not dissolve. Shaking them with water creates a suspension. Larger grains quickly settle to the bottom. Suspensions are not clear and you cannot easily see through them.

Investigating milk

Is milk a solution or a suspension? Find out by adding just one or two drops of milk to a glass of water. Look closely as the milk falls through the water.

YOU WILL NEED 5
- A TALL, CLEAR PLASTIC TUMBLER OF WATER
- MILK
- A TEASPOON

Is milk a single substance or a mixture?

What's going on?

You cannot see clearly through milk, even when you add it to water. Milk consists of droplets of fat suspended in water. Fat is insoluble in water and the droplets are too small for them to settle. Scientists call mixtures like milk 'emulsions'. This name is also given to emulsion paint, which consists of microscopic coloured droplets of oil suspended in water.

FLASHBACK

Creating colours

Paint is coloured because it contains tiny coloured particles called pigment. These particles are suspended in a liquid called the binder that sets hard when exposed to the air. Early artists ground up coloured minerals or chemicals to make pigments. To bind the colour, they used egg white or sticky oils made from boiled tree sap.

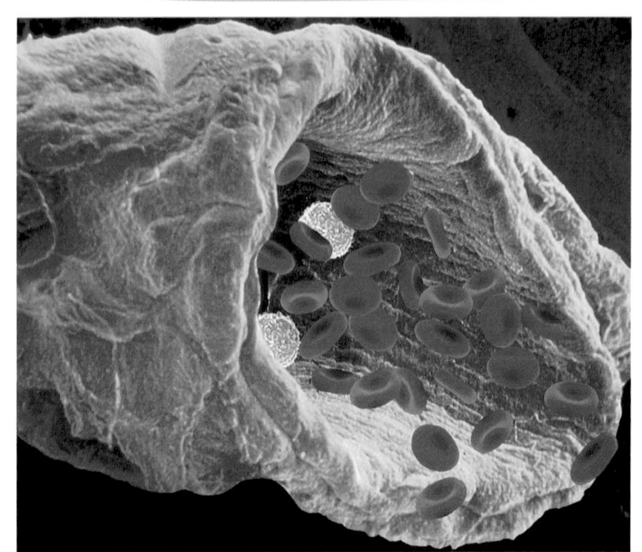

A MIX OF MIXTURES
Blood is both a solution and a suspension. It flows around your body in tubes called veins and arteries. Solid red and white blood cells are suspended in a clear liquid called plasma. This liquid is a solution of hundreds of different substances dissolved in water.

Filtering mixtures

Muddy water is an example of a suspension. It consists of tiny solid particles scattered through a liquid. To separate the particles from the suspension, you can use a filter. Filters work like sieves, but they have microscopic holes called pores, and are often made from thick, fluffy paper. The liquid part of a suspension passes through the holes between the paper fibres, while the solid particles are trapped.

Filtering flour

Mixing flour with water makes a cloudy suspension. Coffee filter paper makes the water clear again.

YOU WILL NEED

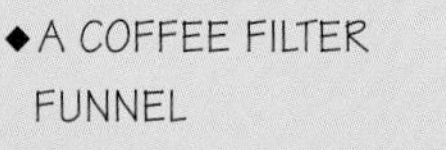

- A COFFEE FILTER FUNNEL
- COFFEE FILTER PAPER
- WATER
- PLAIN FLOUR
- THREE CLEAR PLASTIC TUMBLERS
- A TEASPOON

Which tumbler has the clearest liquid?

1 Add half a teaspoonful of flour to one of the tumblers. Fill the tumbler with water and stir the mixture to make a suspension of flour in water.

2 Place the funnel inside an empty tumbler and put a filter paper inside the funnel. Pour two thirds of the flour and water mixture into the filter.

3 When the tumbler is about a third full, move the funnel and the filter on to the last empty tumbler. Look inside the filter paper when all the liquid has run through. Now look at the liquid in each tumbler and notice the difference.

What's going on?

At first, liquid runs quickly through the filter. Most solid particles are trapped, but some small particles pass through. As a result, the filtered liquid in the first tumbler is slightly hazy. Liquid then passes slowly as the filter pores get blocked. Now even very small particles cannot pass, so the filtered liquid in the third tumbler is almost clear.

Filtering through sand

Cut the bottle in half. Place the funnelled end facing downwards into the base of the bottle. Fill the bottle with cotton wool, pebbles, gravel and sand, as shown, to make your filter. Pour compost mixed with water into the bottle and watch it drip through. What colour are the drips? How fast is the water passing through?

YOU WILL NEED 15

- A 500ML PLASTIC DRINKS BOTTLE
- SCISSORS
- COTTON WOOL
- SAND, GRAVEL, PEBBLES
- POTTING COMPOST
- WATER

How can filters make our tap water clean?

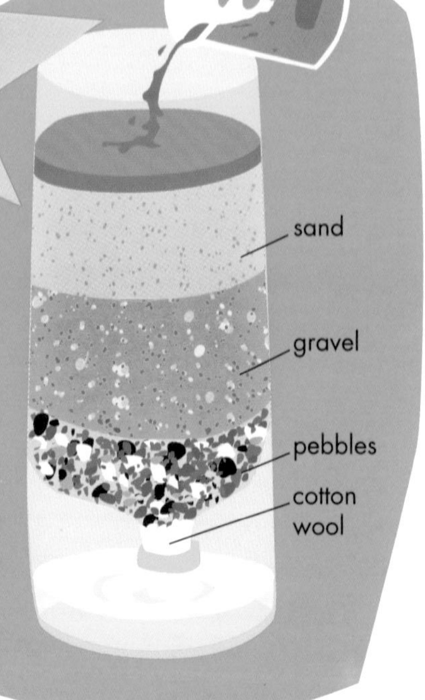

What's going on?

The pebbles, gravel and sand and the fibres in the cotton wool act together as a filter. They prevent the solids in the water from passing through. The trapped solids are called the residue and the liquid that passes through is called the filtrate. The tap water we drink often comes from rivers and lakes. It passes through huge sand filters that make the water clear and pure. Added chemicals kill germs.

FLASHBACK

Filtering germs

In about 1880, doctors discovered that many diseases are caused by germs. They separated germs into two sorts – 'filterable' and 'non-filterable'. 'Filterable' germs called bacteria cause illnesses such as food poisoning. They are large enough to be trapped by a filter. 'Non-filterable' germs are much smaller and pass through a filter. They are called viruses and they cause diseases like chickenpox and flu.

FEEDING THROUGH FILTERS
The humpback whale is a filter feeder. It has hundreds of thin plates called baleen in its mouth. The inside edges of the plates have brushlike fibres, which filter out food particles from the water. Every time the whale scoops about 4,000 litres of water into its mouth, it filters 20kg of tiny food particles.

Evaporating solutions

You can make a solution by dissolving a solid such as salt in water. The solution looks like pure water because the solid has broken down into tiny invisible particles. To make the solid reappear, you can cause the solution to evaporate, or turn into a gas. As the liquid disappears, the solid reappears because there is not enough liquid to dissolve it.

Evaporating salt solution

Solid salt seems to disappear when it dissolves in water. You can evaporate the water to get the solid salt back again.

YOU WILL NEED

20

- SALT
- WARM WATER
- A SAUCER
- A CLEAR PLASTIC TUMBLER
- A TEASPOON

What makes water evaporate?

1 Pour warm water into the tumbler until it is one third full. Add a spoonful of salt and stir until all the salt has dissolved.

2 Pour the salt solution into the saucer until there is a shallow pool, then put the saucer on a sunny windowsill or in some other warm, airy place.

3 Check the saucer twice a day for the next two or three days. What do you notice appearing on the saucer as the water gradually disappears?

What's going on?

Heat causes the water to evaporate – it changes into an invisible gas called water vapour, which escapes into the air. As the liquid slowly evaporates from the solution into the air, the dissolved salt stays behind. You will see a crusty layer of solid salt left on the saucer once all the water has evaporated.

Stalactite on a string

Fill each jar up to three quarters with hot water, then stir in sugar until no more dissolves. Fix a paperclip to each end of the wool. Drop each end into a jar so the wool hangs down between the jars. Place a saucer between the jars and leave them in a warm place. Inspect the string every day for about a week.

YOU WILL NEED 15

- A LENGTH OF WOOL
- TWO PAPERCLIPS
- HAND-HOT WATER
- A DISH
- A SPOON
- TWO JARS
- SUGAR

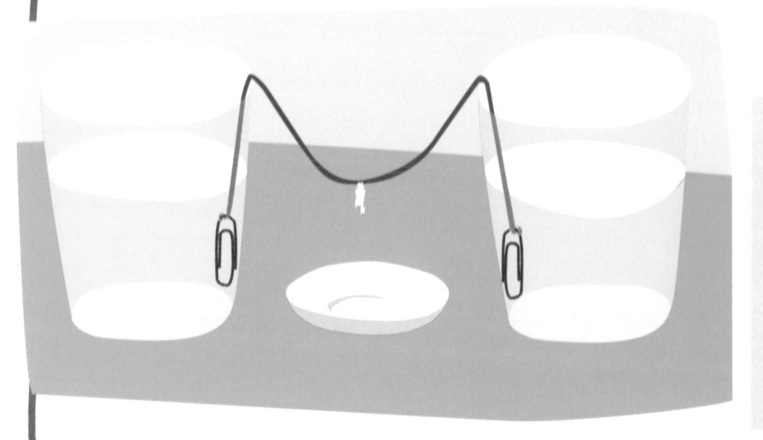

What's going on?

The solution in each jar is saturated – it is as full as it can be of dissolved sugar. The liquid soaks along the wool and collects at the lowest point between the jars. Water evaporates here, so the solid cannot remain dissolved. Solid sugar crystals form and grow bigger as the wool soaks up more solution from the jars.

FLASHBACK

Sea salt

The sea contains many dissolved substances, especially the salt that we use in our food. For thousands of years, people have used heat from the Sun to make solid salt by evaporating sea water. The Latin word for salt is 'sal', which is the origin of the word 'salary'. Roman soldiers were given salt as part of their wages.

STALACTITES AND STALAGMITES
Rainwater often trickles through underground cracks, dissolving the limestone rock underneath. As the water drips from the roof of a cave, it evaporates to leave deposits of solid limestone. Over thousands of years, they grows downwards to become stalactites. Where the drips land, stalagmites grow up from the floor.

Saturated solutions

How much sugar can you dissolve in a cup of coffee? The answer is about 20 spoonfuls. If you add any more, solid sugar stays undissolved in the bottom of the cup. When a solution cannot dissolve any more solid, it is called a saturated solution. The amount of solid needed to make a saturated solution varies from one substance to another.

How much solid?

The solubility of a substance is the amount needed to make a saturated solution. Different substances have different solubilities.

Which substance is the most soluble?

YOU WILL NEED 25

- BICARBONATE OF SODA
- SALT
- SUGAR
- SIX TEASPOONS
- THREE CLEAR PLASTIC TUMBLERS
- WATER
- STICKY LABELS AND PEN

1 Label each of the tumblers 'sugar', 'salt' etc. Half fill them with water and place a teaspoon in each.

2 Add a teaspoonful of sugar to the tumbler labelled 'sugar'. Stir until the solid has dissolved. Now repeat this step in the other tumblers using bicarbonate of soda and salt.

3 Add more solid to each tumbler until no more will dissolve. Count how many spoonfuls of solid dissolve in each tumbler.

What's going on?

The tumblers each contain the same amount of water to make sure the test is fair. More sugar dissolves than salt, so you can say that sugar is more soluble than salt. Less bicarbonate of soda dissolves than sugar or salt, so it is the least soluble of the three substances.

Growing crystals

Half fill a tumbler with warm water. Stir in sugar until no more will dissolve, then pour the clear solution into the other tumbler, leaving any undisolved sugar behind. Use the pencil and cotton to suspend the paperclip in the solution. Look at the paperclip every day for about a week and see what happens.

YOU WILL NEED

- SUGAR, (OR WASHING SODA)
- TWO CLEAR PLASTIC TUMBLERS
- A PENCIL
- COTTON
- A PAPERCLIP

Where do most of the crystals grow?

What's going on?

The water slowly evaporates and crystals appear when there is not enough water to dissolve all the solid. Crystals grow on places that aren't smooth, so you'll see them first on the edges of the paperclip. The water in the tumbler disappears slowly because there is only a small surface area from which it can evaporate. This slow evaporation helps large crystals to grow.

Fizzing bubbles

Place one bottle of fizzy drink in the fridge and the other in the bucket of warm water. Half an hour later, open both the bottles (over the sink!) What do you see?

YOU WILL NEED

5

- TWO SMALL BOTTLES OF FIZZY DRINK
- FRIDGE
- A BUCKET OF WARM WATER

Which drink – cool or warm – froths the most?

What's going on?

Fizzy drinks consist of carbon dioxide gas dissolved in flavoured water. You will see more of this gas froth out from the warm drink than from the cold drink. This is because more gas can be dissolved in cold liquids than in warm liquids. As you open the bottle, the pressure inside is released, which allows gas to bubble out of the solution and escape.

SUGAR SEEDS

Sugar is made from the juice of sugar cane and sugar beet. One hundred grams of tiny seed crystals are added to a huge tank filled with a saturated solution of sugary syrup. It takes only two hours for each seed to grow until the tank contains 20 tonnes of solid sugar crystals.

Glossary

Boil When bubbles quickly grow in a liquid, rise to the surface and burst, releasing vapour. Boiling is the most rapid form of evaporation.

Brittle Describing solids that snap easily when bent or shatter into pieces when struck. The opposite of tough.

Chemical A single, pure substance. Salt is a chemical that chemists call sodium chloride.

Chemist A scientist who studies how permanent changes can make new substances. Remember that the people in charge of chemists's shops are called pharmacists. They prescribe medicines and drugs.

Combustion Another word for burning.

Compress To squeeze something so that its volume decreases and it takes up less room than it had before. It is fairly easy to compress gases. It is almost impossible to compress liquids or solids.

Condense To change a gas into a liquid, usually by cooling it.

Conductor A solid that allows heat (and electricity) to pass easily through it. Metals such as copper and aluminium are good conductors (*see* **Insulator**).

Contract When an object becomes smaller. Most solids and all liquids and gases contract when they cool and their temperatures decrease.

Dissolve When a substance disappears as it mixes into a liquid. Salt dissolves in water to make salt solution.

Elastic A solid that changes shape when squeezed or stretched; it then returns to its original shape when the squeezing or stretching stops.

Energy Energy is needed to make things happen. It is the ability to do work. Heat and electricity are two types of energy. Fuels contain energy that is released as heat when they burn.

Evaporate When a liquid changes into a vapour (gas), usually by heating it.

Expand When an object becomes larger. Solids, liquids or gases expand when they are heated and their temperatures increase.

Filtrate The liquid part of a suspension that passes through a filter.

Force A push or a pull. Force can do work and make things speed up, slow down, or change shape. Forces can also cancel when they push or pull against each other.

Freeze When a liquid changes into a solid, usually by cooling it.

Heat A form of energy. When heat flows into an object, its temperature increases. The temperature decreases when heat flows out of an object.

Insoluble A substance that does not dissolve in a liquid.

Insulator A substance that does not allow heat (and electricity) to pass easily through it. Most liquids and gases and solids such as wood and plastic are insulators.

Length A measurement of the distance

between two places. The unit of length is the metre (m). One metre equals 100 centimetres (cm) or 1000 millimetres (mm). One kilometre (km) is equal to 1000m.

Mass The amount of matter in an object. The unit of mass is the kilogram (kg). One kilogram is equal to 1000 grams (g). One thousand kilograms is equal to 1 tonne.

Material Different kinds of solids. Steel, paper, skin, stone and plastic are all materials. Objects are made by fitting different materials together.

Matter Anything that has mass and takes up space.

Melt When a solid changes into a liquid, usually by heating it.

Permanent Describing a change that cannot easily be reversed.

Pressure A measurement of the amount of force pressing on the surface of an object. Your feet exert pressure on the floor. The pressure of the air inside a balloon keeps the skin stretched outwards.

Raw materials Natural substances that are used to make useful products. Raw materials are extracted from the ground (e.g. iron ore, crude oil), from seawater (e.g. bromine and iodine for use in medicines) and from the air (e.g. oxygen and nitrogen).

Saturated solution A solution that cannot dissolve any more solid.

Solidify When a liquid changes into a solid, usually by cooling it.

Solubility A measurement of how much solid (or gas) dissolves in a fixed amount of liquid.

Soluble A substance that will dissolve in a liquid.

Solution The mixture that results when a substance dissolves in a liquid.

Substance Any kind of matter. A substance can be a solid, a liquid, or a gas. A common word for substance is 'stuff'.

Suspension A mixture made by shaking small insoluble particles with a liquid.

Temperature A measurement that describes how hot something is. On the Celsius temperature scale, water freezes at 0°C and boils at 100°C.

Temporary Describing a change that can easily be reversed.

Tough Solids that do not bend easily and that do not break into pieces when struck. The opposite of brittle.

Vapour Another word for gas.

Volume A measurement of the amount of space taken up by an object. The unit of volume is the litre (l). One litre is equal to 1000 millilitres (ml). Millilitres are sometimes called cubic centimetres (cm^3).

Weight The force on an object that results when gravity pulls on its mass. A bag of sugar has a mass of 1kg on Earth and 1kg on the Moon. Its weight on Earth is six times its weight on the Moon because gravity on Earth is six times stronger than on the Moon.

Index

Picture Credits
Bruce Coleman (bottom right) 13, 33
DIY Photo Library (bottom right) 17
Impact Photos (bottom right) 21
Inter IKEA Systems B.V.1999 (bottom right) 7
Photofusion (bottom right) 25
Rex Features (bottom right) 27
Robert Harding (bottom right) 9, 11, 15, 23, 31, 35
Science Photo Library (bottom right) 37
Tony Stone (bottom right) 19
The Stock Market (bottom right) 29